IMPROVING WORK THROUGH ERGONOMICS IN THE DIGITAL AGE

Paul Jos

Table of Contents

1. Preface .... 4.

2. **Introduction to Ergonomics**: An overview of the field, its history, and its importance in designing efficient and safe work environments .... 6.

3. **Human Factors and Human-Machine Interaction**: Understanding how humans interact with machines, tools, and technology .... 8.

4. **Anthropometry**: The study of human body measurements and proportions, crucial for designing products and spaces that fit people comfortably ... 13.

5. **Biomechanics**: Examining how forces and movements affect the human body during work and daily activities .... 19.

6. **Workplace Design and Layout**: Strategies for creating ergonomic workstations, including seating, lighting, and equipment placement .... 23.

7. **Cognitive Ergonomics**: Investigating mental processes, decision-making, and information processing in work settings .... 27.

8. **Environmental Ergonomics**: Addressing factors like temperature, noise, and air quality that impact human performance .... 30.

9. **Musculoskeletal Disorders and Injury Prevention**: Identifying risks and implementing measures to prevent workplace injuries .... 34.

10. **Ergonomics in Product Design**: Applying ergonomic principles to design consumer products, tools, and gadgets .... 37.

11. **Ergonomics in Healthcare and Special Populations**: Focusing on healthcare settings, aging populations, and individuals with disabilities .... 43.

12. Enhancing Ergonomics with the Assistance Exoskeletons .... 49.

13. Leveraging AI (artificial intelligence) for Ergonomics .... 51.

14. Ergonomics Training .... 55.

15. Ergonomics and Remote work .... 58.

16. Strategy for Answering Ergonomics Questions...60.

17. Self-Study Questions on Each Chapter..63.

18. Epilogue...79.

19. Bibliograph..81.

Preface

Welcome to the fascinating world of ergonomics! In this book, we delve into the intricate balance between human capabilities, environmental factors, and efficient system design. Our journey will explore how ergonomics plays a pivotal role in creating work environments that enhance both productivity and safety.

Why Ergonomics Matters

Ergonomics is more than just a buzzword; it's a science that shapes our daily lives. As you turn the pages, you'll discover how ergonomics influences everything from the layout of workstations to the design of consumer products. Whether you're an engineer, a safety professional, or simply curious about human factors, this book offers valuable insights.

What Awaits You

1. After this summary "preface"
2. Introduction to Ergonomics: We kick off with an overview of the field's history and its significance in *designing efficient and safe workspaces*. You'll gain a deeper understanding of why ergonomics matters.
3. Human Factors and Human-Machine Interaction: Explore how *humans interact with machines, tools, and technology*. From intuitive interfaces to user-friendly gadgets, we unravel the secrets behind seamless interactions.
4. Anthropometry: Dive into the study of *human body measurements and proportions*. Learn why designing products and spaces that fit people comfortably is essential.
5. Biomechanics: Understand *how forces and movements impact the human body* during work and daily activities. Discover the science behind ergonomic design.
6. Workplace Design and Layout: *Strategies for creating ergonomic workstations* await you. From seating arrangements to optimal lighting, we'll guide you through the essentials.
7. Cognitive Ergonomics: Peek into *mental processes, decision-making, and information processing in work settings*. Uncover the psychology behind efficient workflows.
8. Environmental Ergonomics: *Temperature, noise, and air quality*—all factors that influence human performance—are explored in this section.
9. *Musculoskeletal Disorders and Injury Prevention*: Identifying risks and implementing preventive measures is crucial. Learn how to keep workplaces safe.

10. Ergonomics in Product Design: From smartphones to power tools, we apply ergonomic principles to consumer products. *Discover the art of user-centered design.*
11. Ergonomics in *Healthcare and Special Populations*: Healthcare settings, aging populations, and individuals with disabilities—all covered here.
12. Enhancing Ergonomics with Assistance Exoskeletons: *Cutting-edge technology meets ergonomic solutions.* Explore the future of work.
13. *Leveraging AI for Ergonomics*: Artificial intelligence isn't just science fiction—it's a powerful tool for optimizing ergonomic designs.
14. Ergonomics Training: Equip yourself with practical knowledge. *Training matters*, and we'll show you why.
15. Ergonomics and Remote Work: *In an increasingly digital world, remote work presents unique challenges.* Let's address them.
16. Strategy for Answering Ergonomics Questions: *A practical guide for tackling real-world problems.*
17. Chapters Self-Study Questions: *Test your understanding and reinforce key concepts.*

As you embark on this ergonomic journey, remember that every design choice impacts human well-being. So, let's create workspaces that harmonize efficiency, safety, and comfort!

—Paul Jos

CHAPTER ONE

Introduction to Ergonomics

What is Ergonomics?

Ergonomics, also known as human factors engineering, is the scientific discipline that focuses on designing and arranging work environments, products, and systems to optimize human performance, safety, and well-being. The word "ergonomics" is derived from the Greek words "ergon" (meaning work) and "nomos" (meaning laws or rules). Essentially, ergonomics aims to establish rules and guidelines for creating workspaces and tools that fit human capabilities and limitations.

Historical Context

The roots of ergonomics trace back to ancient civilizations, where craftsmen and builders intuitively adjusted their tools and workspaces to enhance efficiency and comfort. However, the formal study of ergonomics began during the Industrial Revolution in the 19th century. As factories and assembly lines emerged, researchers and engineers recognized the need to improve worker productivity and reduce injuries.

One of the earliest pioneers of ergonomics was *Frederick Winslow Taylor*, an American engineer who introduced scientific management principles. Taylor's work emphasized time-motion studies, standardization, and efficiency. While his focus was primarily on productivity, it laid the groundwork for ergonomic principles.

Key Concepts in Ergonomics

1. Human-Centered Design

Ergonomics places humans at the center of design. It acknowledges that people have varying *physical abilities, cognitive capacities, and sensory perceptions*. Designing with the user in mind ensures that products and workspaces accommodate these differences.

2. Fit and Comfort

An ergonomic design aims for a "good fit" between the user and the environment. This includes proper chair height, keyboard placement, and lighting. Comfort is essential for sustained productivity and well-being.

3. Safety and Injury Prevention

Ergonomics seeks to prevent work-related injuries and musculoskeletal disorders (such as back pain, carpal tunnel syndrome, and repetitive strain injuries). Proper workstation setup, lifting techniques, and posture play critical roles in injury prevention.

4. Productivity and Efficiency

Efficient work processes lead to higher productivity. Ergonomic designs optimize workflow, reduce unnecessary movements, and enhance task performance.

5. Multidisciplinary Approach

Ergonomics draws knowledge from various fields, including engineering, psychology, physiology, and design. Collaboration among experts ensures comprehensive solutions.

Importance of Ergonomics

1. Health and Well-Being

Ergonomics directly impacts physical health and overall well-being. A well-designed workspace reduces the risk of strain, fatigue, and discomfort.

2. Productivity and Quality

Efficient work environments enhance productivity and product quality. Employees perform better when their tools and surroundings support their needs.

3. Cost Savings

Investing in ergonomic design pays off in the long run. Reduced absenteeism, fewer workplace injuries, and increased efficiency lead to cost savings for organizations.

Overall, ergonomics is not a luxury; it is a necessity. As our work environments evolve, *understanding and applying ergonomic principles become increasingly crucial.* By prioritizing human factors, we create safer, healthier, and more efficient workplaces.

CHAPTER TWO

Human Factors and Human-Machine Interaction

Understanding How Humans Interact with Machines, Tools, and Technology

In this chapter, we delve into the fascinating realm of human factors and human-machine interaction (HMI). These fields explore the intricate relationship between humans and the technology they use. Let's unravel the core concepts and principles that shape our interactions with machines.

1. Human Factors: The Science of Interaction

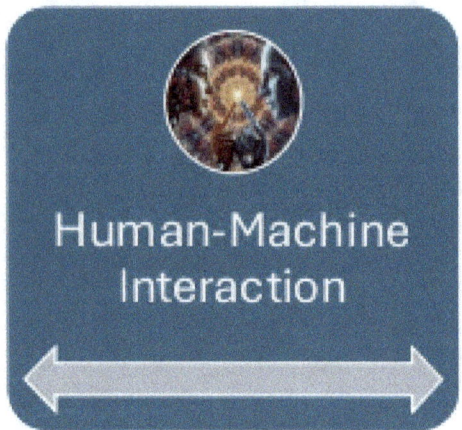

What are Human Factors?

Human factors, also known as ergonomics, is the scientific discipline that studies how people interact with devices, products, and systems. It's an applied field where behavioral science, engineering, and other disciplines converge to develop principles that ensure devices and systems are usable by their intended users.

Key Aspects of Human Factors:

- Usability: Ensuring that products and interfaces are intuitive, efficient, and user-friendly.

- Safety: Designing systems that minimize errors and prevent accidents.

- User Experience (UX): Focusing on how users perceive and interact with technology.

2. Human-Machine Interaction: Bridging the Gap

Defining HMI:

Human-machine interaction explores the dynamics between humans and machines. It encompasses various forms of interaction, including visual, auditory, and physical exchanges. Let's break it down:

- Visual Interaction: How we perceive information through screens, displays, and visual cues.

- Auditory Interaction: How sound and speech influence our interactions with technology.

- Physical Interaction: Touchscreens, buttons, and gestures—our tactile engagement with devices.

3. Challenges and Considerations

Levels of Autonomy:

- Balancing control between humans and machines.

- From manual control to full automation—finding the sweet spot.

Teaming:

- Collaborative, cooperative, co-active, or competitive interactions.

- How teamwork affects performance and decision-making.

Human Performance Enhancement:

- Designing technology to augment human capabilities.

- Enhancing productivity, accuracy, and efficiency.

Shared Control:

- Striking a balance between machine autonomy and human oversight.

- Ensuring safety while empowering users.

4. Formal Description Techniques for Interfaces

Model-Based Approaches:

- Using formal models to design and evaluate interactive systems.

- Ensuring dependability and usability.

Designing Human-Automation Interaction:

- Crafting interfaces that seamlessly integrate human and machine.

- Addressing cognitive load, decision support, and adaptability.

Human-Agent Interaction:

- How humans interact with intelligent agents (e.g., chatbots, virtual assistants).

- Trust, transparency, and effective communication.

5. Future Directions

Personalized Interaction:

- Tailoring interfaces to individual needs and preferences.

- Adaptive systems that learn from user behavior.

Clinical Implementations:

- Applying HMI principles in healthcare and rehabilitation.

- Enhancing patient outcomes through personalized interactions.

Practical Tips and Applications

As technology evolves, understanding human-machine interaction becomes paramount. By designing interfaces that align with our cognitive abilities, emotions, and physical capabilities, we create a world where machines enhance our lives rather than hinder them and thereby provide a positive user experience. Here are some practical tips on how to enhance the usability of software.

1. Easy Learning Curve:

   - Ensure that your program or software is easy to learn. Users should be able to grasp the basics without complex rules or manuals.

   - Intuitive interfaces, like Facebook's, contribute to popularity because they are straightforward and user-friendly.

2. Informative Home Page:

   - Design a home page that is informative and presents multiple options clearly.

   - Users should understand where each option leads without confusion.

3. Simplified Navigation:

   - Basic navigation should be simple and efficient.

   - Users should find what they need quickly, avoiding frustration and negative impressions.

4. Efficiency Matters:

   - Prioritize efficiency in your software. Users appreciate streamlined processes.

   - Optimize workflows, minimize unnecessary steps, and reduce friction.

5. Memorability:

   - Make your software easy to memorize. Users should remember how to operate it even after a break.

   - Consistent layouts and clear icons aid memorability.

6. Error Prevention and Guidelines:

   - Implement proper guidelines and double-check measures to prevent mistakes.

   - Error messages should be informative and guide users toward resolution.

7. User Satisfaction:

   - Ultimately, usability is about ensuring users are satisfied with their experience.

   - Regularly gather feedback and iterate based on user needs and preferences.

8. Unique Features:

   - Consider adding unique features that set your software apart.

   - These features can enhance usability and attract users.

Remember that usability design is an ongoing process. *Always one should* continuously seek feedback, test your software/device, and adapt based on user behavior and preferences.

Examples of human-machine interaction across various domains:

- Internet of Things (IoT) Devices: IoT technology has significantly impacted our daily lives. Smart home devices, wearable health trackers, and connected appliances seamlessly interact with users, enhancing convenience and efficiency.
- Eye-Tracking Technology: Eye-tracking systems detect where a person is looking based on their gaze point. These technologies are used in fields like usability testing, gaming, and medical research.
- Speech Recognition Technology: Speech recognition interprets human language, derives meaning from it, and performs tasks for the user. Virtual assistants like Siri, Google Assistant, and Amazon Alexa rely on this technology.
- Augmented Reality (AR) and Virtual Reality (VR): AR and VR technologies create immersive experiences by blending digital content with the real world (AR) or simulating entirely virtual environments (VR). Applications range from gaming to training simulations.
- Cloud Computing: Cloud services allow seamless data storage, collaboration, and access across devices. Users interact with cloud-based applications, databases, and services without worrying about physical infrastructure.
- Human-Computer Interaction (HCI) in Aviation: Pilots interact with complex cockpit systems, navigation displays, and autopilots during flight. Effective HCI design ensures safe and efficient operation of aircraft.
- Healthcare Diagnostic Programs: Physicians use diagnostic software to analyze medical images (such as X-rays or MRIs) and assist in diagnosing conditions. These tools enhance accuracy and speed in healthcare settings.
- Industrial Machine Operation: Workers operate machinery in manufacturing, construction, and other industries. Well-designed interfaces ensure smooth interaction between humans and machines, improving productivity and safety.
- Automated Driving Systems: Self-driving cars rely on human-machine interaction. Users interact with navigation interfaces, monitor sensor data, and occasionally take control when needed.
- Human-Agent Interaction: Chatbots, virtual assistants, and AI-driven customer service agents engage in conversations with users. Effective interaction design ensures seamless communication and problem-solving.

These examples demonstrate how thoughtful design and collaboration between humans and machines lead to successful outcomes across diverse contexts.

CHAPTER THREE

Anthropometry: Designing for Human Proportions

Understanding Human Body Measurements

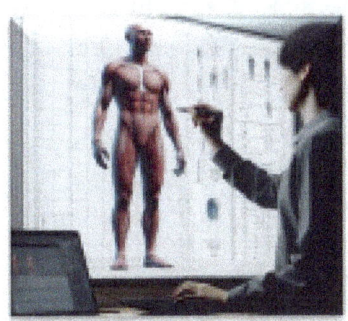

Anthropometry is the scientific study of human body measurements, proportions, and physical characteristics. It plays a vital role in designing products, spaces, and interfaces that accommodate the diversity of human shapes and sizes. Let's delve into the significance of anthropometry and its practical applications.

1. Historical Context

Origins of Anthropometry

- Anthropometry dates back centuries, with ancient civilizations using body measurements for clothing, furniture, and architecture.

- Leonardo da Vinci's detailed anatomical sketches exemplify early anthropometric observations.

2. Key Concepts in Anthropometry

Body Dimensions

- Anthropometry encompasses various body dimensions, including:

  - Stature (Height): Vertical measurement from head to toe.

  - Sitting Height: Measurement from the seat surface to the top of the head when sitting.

  - Reach and Arm Length: Crucial for designing workspaces, tools, and vehicles.

- Leg Length: Influences chair and desk height.

- Circumferences: Waist, hip, chest, and limb circumferences.

Proportions

- Anthropometric ratios guide design decisions:

  - Golden Ratio: Aesthetic proportion often seen in architecture and art.

  - Body Segment Ratios: For ergonomic chair design, table height, and clothing fit.

3. Applications in Design

Ergonomics and Workspace Design

- Proper desk and chair heights enhance comfort and productivity.

- Anthropometric data informs workstation dimensions and layout.

Product Design

- Furniture, clothing, and footwear must fit a wide range of body sizes.

- Anthropometric databases aid designers in creating universally comfortable products.

Vehicle Design

- Car interiors, airplane seats, and bicycle frames consider legroom, head clearance, and reach.

- Anthropometric studies optimize driver and passenger comfort.

Architecture and Interior Space

- Doorway heights, stair risers, and room dimensions align with human proportions.

- Public spaces accommodate diverse populations.

4. Challenges and Considerations

Population Diversity

- Anthropometric data varies across age, gender, ethnicity, and geographical regions.

- Designers must account for this diversity.

Dynamic Anthropometry

- Body dimensions change over time (e.g., growth, aging, weight gain).

- Adjustable furniture and adaptable spaces address dynamic needs.

5. Future Trends

Digital Anthropometry

- 3D scanning and virtual modeling enable personalized design.

- Custom-fit clothing and medical devices benefit from precise measurements.

Inclusive Design

- Prioritizing accessibility and inclusivity.

- Designing for people with disabilities or unique requirements.

Practical Tips and Applications

Anthropometry bridges science and design, ensuring that our built environment harmonizes with human proportions. By embracing this multidisciplinary field, we create spaces and products that not only function well but also enhance our well-being and comfort. Applying anthropometric principles to your workspace is essential for creating a comfortable and efficient environment.

Let's explore how you can do this:

Understand Anthropometry:

Because anthropometry involves measuring human body dimensions. It helps ensure that workspaces, equipment, and furniture fit the human body.

Key measurements include eye height, elbow height, hip breadth, and overall stature.

Design Choices:

When applying anthropometric data, consider three basic design choices:

Design for the Average: Use average measurements (50th percentile) to create standard designs.

Design for Adjustability: Allow customization (e.g., adjustable chairs, desks) to accommodate different body sizes.

Design for Extremes: Consider the 5th percentile (small body) and 95th percentile (tall body) for outliers.

Scope Your Design:

Determine which body dimensions are relevant to your workspace.

Consider the function of your product (e.g., chair, desk) and the tasks performed.

Choose Percentiles:

Anthropometric databases categorize measurements into percentiles (e.g., 5th, 50th, 95th).

Decide which percentile(s) you'll accommodate based on your user group.

Data and Allowances:

Obtain anthropometric data from reliable sources (e.g., Army Anthropometric Survey Database, CAESAR).

Add allowances for clothing, footwear, and other worn equipment (e.g., personal protective gear).

Apply to Workspaces:

Desk and chair heights: Optimize for elbow height and knee height.

Monitor placement: Adjust screen height and distance.

Seating depth: Consider leg segment measurements.

Evaluate and Iterate:

Regularly assess the fitness and comfort of your workspace.

Gather user feedback and adjust as needed.

A well-designed workspace enhances productivity, reduces fatigue, and promotes overall well-being.

The following is a typical workspace example for properly adjusting a chair- crucial for maintaining good ergonomics and preventing discomfort: **A step-by-step guide to achieve better chair ergonomics:**

Seat Height:

Adjust the chair height so that your feet rest flat on the floor or on a footrest. Your knees should form a 90-degree angle or slightly lower than your hips.

Seat Depth:

Ensure there is a full fist's space between the back of your knees and the seat pan.

Push the seat pan as far back as possible to avoid gaps between the seat and the backrest.

Backrest Height:

Adjust the backrest height so that the lumbar support aligns with your belt level.

Proper lumbar support stabilizes your pelvis and maintains a natural curve in your lower back.

Armrests:

Position the armrests just below the level of your elbows.

They should not hinder your movement while working.

Headrest (if applicable):

Adjust the headrest to support your head when you lean back and relax.

However, during work, it should not interfere with your posture.

Backrest Angle and Tilt Mechanism:

Set the backrest angle to keep the chair stable while working.

Depending on your habits and chair quality, you can lock or unlock the tilt mechanism.

Desk Height:

If your desk is too high, consider adjusting it or using a footrest.

The footrest should be flat, horizontal, and allow your heels to be under your knees.

One should periodically reassess chair adjustments and make any necessary changes to maintain optimal ergonomics.

CHAPTER FOUR

Biomechanics: Understanding Human Movement and Forces

Unraveling the Mechanics of Our Bodies

Biomechanics is the scientific study of how forces and movements impact the human body. By examining the interplay between physics, anatomy, and physiology, we gain insights into how our bodies function during work, exercise, and everyday tasks. Let's delve into this fascinating field.

1. The Basics of Biomechanics

Forces and Motion

- Biomechanics investigates how external forces (gravity, friction, muscle tension) influence our bodies.

- Newton's laws of motion guide our understanding of movement.

Kinematics and Kinetics

- Kinematics studies motion patterns (position, velocity, acceleration).

- Kinetics delves into the forces causing motion (muscle contractions, joint reactions).

2. Applications in Daily Life

Walking and Running

- Biomechanics explains the efficiency of our gait.

- Stride length, foot strike, and joint angles impact our locomotion.

Lifting and Carrying

- Proper lifting techniques prevent strain and injury.

- Biomechanical analysis guides safe lifting practices.

Sports Performance

- Athletes optimize movements based on biomechanical principles.

- Tennis serves, golf swings, and soccer kicks all rely on precise mechanics.

3. Ergonomics and Workplace Design

Workplace Biomechanics

- Ergonomic chairs, desks, and tools align with natural body movements.

- Reducing strain improves productivity and well-being.

Manual Material Handling

- Biomechanics informs lifting, pushing, and pulling techniques.

- Proper body mechanics prevent musculoskeletal injuries.

4. Injury Prevention and Rehabilitation

Understanding Injury Mechanisms

- Biomechanics identifies risk factors for injuries (e.g., ACL tears, back pain).

- Sports scientists and physical therapists apply this knowledge.

Prosthetics and Orthotics

- Designing functional artificial limbs relies on biomechanical principles.

- Orthotic devices support injured joints and aid recovery.

5. Future Frontiers

Biomechanics in Virtual Reality

- Simulating movement and assessing biomechanics in virtual environments.

- Training athletes and rehabilitating patients using VR.

Bioinspired Design

- Drawing inspiration from nature (e.g., animal locomotion, plant structures).

- Creating innovative materials and devices.

Practical Tips and Applications:

Biomechanics bridges science and practicality, enhancing our understanding of how our bodies navigate the world. Whether you're an athlete, office worker, or curious explorer, appreciating biomechanics enriches your daily experiences. Biomechanics plays a pivotal role in advancing medical research and practice. Here are some cutting-edge applications of biomechanics in medicine:

Cell and Gene Therapies:

Biomechanics informs the design of delivery systems for cell-based therapies. Understanding how forces affect cell behavior helps optimize treatments.

Antiaging Treatments:

Biomechanics contributes to innovations in antiaging therapies. Researchers explore tissue mechanics, collagen regeneration, and skin elasticity.

Drug Delivery Systems:

Biomechanical principles guide the development of drug delivery devices. These systems ensure precise drug release and minimize tissue damage.

Predictive Modeling of Health and Disease:

Computational biomechanics models predict disease progression and evaluate treatment outcomes. For example, modeling blood flow in arteries aids in diagnosing cardiovascular diseases.

3D-Printed Biocompatible Scaffolds:

Biomechanics-driven design produces patient-specific scaffolds for tissue engineering and regenerative medicine.

Microelectronic Devices:

Implantable sensors and microdevices monitor physiological parameters. Biomechanical compatibility ensures long-term functionality.

Corneal Biomechanics:

Clinical devices measure and modify corneal biomechanics in vivo. These advancements enhance diagnostics and personalized treatments for eye conditions.

Rehabilitation Technologies:

Biomechanics guides the development of prosthetics and orthotics. Custom-fit devices improve mobility and quality of life for amputees.

Deep Learning and Diagnostic Tools:

Biomechanical data combined with deep learning algorithms enhances disease detection and risk assessment. For instance, analyzing gait patterns aids in diagnosing neurodegenerative disorders.

Bioinspired Design:

Researchers draw inspiration from natural biomechanics to create innovative medical solutions. Mimicking biological structures leads to novel materials and devices.

These applications demonstrate how biomechanics contributes to personalized medicine, improved treatments, and better patient outcomes.

CHAPTER FIVE

Workplace Design and Layout: Creating Ergonomic Workstations

Designing Workspaces for Comfort and Productivity

In this chapter, we explore strategies for designing ergonomic workstations that promote well-being, productivity, and overall satisfaction. Whether you're setting up an office, home workspace, or industrial environment, thoughtful design plays a crucial role.

1. Ergonomic Principles

Understanding Ergonomics

- Ergonomics focuses on adapting the work environment to fit human capabilities and limitations.

- A well-designed workspace minimizes physical strain and enhances performance.

Seating Considerations

- Chair Selection: Choose adjustable chairs with lumbar support, seat depth, and armrests.

- Seat Height: Adjust the chair height to allow feet to rest flat on the floor or footrest.

Lighting and Visibility

- Natural Light: Position workstations near windows to maximize natural light.

- Task Lighting: Provide adjustable task lighting to reduce eye strain.

2. Workstation Layout

Desk Placement

- Monitor Position: Place the monitor at eye level, about 20 inches from your eyes.

- 23 -

- Keyboard and Mouse: Keep them at elbow height to maintain neutral wrist positions.

Equipment Arrangement

- Frequently Used Items: Arrange them within easy reach to minimize unnecessary movements.

- Phone and Documents: Position them close to avoid excessive stretching or twisting.

3. Computer Workstations

Monitor Ergonomics

- Eye Level: Adjust the monitor height to avoid neck strain.

- Distance: Maintain a comfortable viewing distance to reduce eye fatigue.

Keyboard and Mouse Placement

- Keyboard Tilt: Slightly tilt the keyboard away from you to maintain a neutral wrist position.

- Mouse Position: Keep it close to the keyboard to minimize reaching.

4. Lighting and Color

Lighting Quality

- Even Illumination: Avoid glare and shadows.

- Color Temperature: Opt for neutral or warm lighting.

Color Psychology

- Productivity: Blue and green promote focus.

- Creativity: Yellow and orange encourage creativity.

5. Noise and Privacy

Noise Reduction

- Acoustic Panels: Install sound-absorbing panels to minimize distractions.

- Headphones: Provide noise-canceling headphones for focused work.

Privacy Solutions

- Cubicles or Partitions: Create individual work zones.

- Quiet Rooms: Designate spaces for private calls or concentration.

Practical Tips and Applications

Effective workplace design goes beyond aesthetics—it directly impacts health, efficiency, and job satisfaction. By integrating ergonomic principles, thoughtful layout, and user-friendly features, we create workspaces that empower individuals to thrive. One typical example is implementing proper lighting in your home office which is essential for productivity, comfort, and overall well-being. Here are some tips to enhance the lighting in your workspace:

Diffuse Ambient Light:

Avoid working directly under the glare of overhead lights. Use lampshades to soften and scatter harsh light. Consider an upward-shining floor lamp to bounce light off walls and ceilings, creating a more even illumination without undue glare or shadows.

Task Lighting:

For focused tasks like computer work or paperwork, choose a dedicated light source. An adjustable desk lamp allows precise positioning to illuminate specific areas. If your home office has multiple workstations (e.g., computer desk, filing area, photo review table), set up task lighting for each station.

Avoid Glare and Shadows:

Be mindful of where your light sources are placed. A light behind you while working on the computer can create glare on your monitor. Ensure that task lights don't cast unintended shadows (e.g., writing with your right hand while the light is on the right side).

Leverage Natural Light:

If possible, position your workstation near a window or skylight. Sunlight provides warm lighting that improves the work environment. Be mindful of direct sunlight causing overwhelming glare during certain times of the day. Position your workstation facing north or south to avoid shadows throughout the day. Use solar shades or blinds to soften sunlight without compromising the view. *Note that a well-lit home office positively impacts your work efficiency and overall mood.* Experiment with different lighting options to find what works best for you.

CHAPTER SIX

Cognitive Ergonomics: Enhancing Mental Efficiency in Work Environments

Unraveling the Mind's Workings

Cognitive ergonomics delves into the intricate interplay between human cognition and work settings. By understanding mental processes, decision-making, and information flow, we can optimize productivity, reduce errors, and foster well-being. Let's explore this fascinating field.

1. The Cognitive Landscape

Information Processing

- Cognitive ergonomics examines how our brains process information.

- From perception to memory, understanding these stages informs workplace design.

Attention and Focus

- Designing workspaces that minimize distractions enhances focus.

- Proper lighting, noise control, and layout impact attention span.

2. Decision-Making and Problem-Solving

Heuristics and Biases

- Cognitive shortcuts (heuristics) influence decision-making.

- Awareness of biases (confirmation bias, anchoring) helps mitigate errors.

Decision Support Systems

- Integrating technology (AI, data analytics) aids decision-making.

- Dashboards, predictive models, and real-time feedback enhance cognitive efficiency.

- 27 -

3. Mental Load and Stress

Cognitive Load

- Balancing mental resources during tasks.

- Reducing unnecessary cognitive load improves performance.

Stress Management

- High-stress environments impair cognitive function.

- Strategies like mindfulness, breaks, and relaxation techniques mitigate stress.

4. Human-Computer Interaction (HCI)

User Interface Design

- Intuitive interfaces minimize cognitive effort.

- Consistent layouts, clear icons, and logical workflows enhance usability.

Adaptive Systems

- AI-driven systems adapt to user behavior.

- Personalized recommendations and context-aware interfaces optimize cognitive load.

5. Future Frontiers

Neuro-ergonomics

- Brain-computer interfaces (BCIs) merge cognition and technology.

- Controlling devices with thoughts opens new possibilities.

Collaborative Cognition

- Studying group dynamics and distributed cognition.

- How teams process information collectively.

Practical Tips and Applications

Cognitive ergonomics isn't just about desks and chairs—it's about designing work environments that align with our mental capacities. By optimizing attention, decision-making, and stress management, we create workplaces where minds thrive. Enhancing collaboration in the workplace is essential for productivity, creativity, and team satisfaction. Here are practical strategies to foster effective collaboration:

Build Psychological Safety and Trust:

Create an environment where team members feel safe to express ideas, ask questions, and take risks. Encourage open dialogue and active listening to build trust among colleagues.

Delegate Effectively:

Empower team members by delegating tasks based on their strengths and expertise. Trust your team to take ownership of their responsibilities and contribute effectively.

Implement a Decision-Making Framework:

Establish clear decision-making processes. Whether it's consensus-based, democratic, or hierarchical, consistency ensures everyone understands their role in decision-making.

Promote Open and Transparent Communication:

Encourage honest discussions and constructive feedback. Regular check-ins, team meetings, and transparent communication channels foster collaboration.

Create Professional Development Opportunities:

Invest in training, workshops, and skill-building sessions. When team members grow together, they collaborate more effectively and bring fresh perspectives to the table.

Collaboration shouldn't just be about tools and technology—it should be about creating a culture where individuals thrive together.

CHAPTER SEVEN

Environmental Ergonomics: Optimizing Human Performance

Navigating the Elements for Productivity and Well-Being

Environmental ergonomics delves into the impact of our surroundings on human performance. By addressing factors like temperature, noise, and air quality, we create workspaces that enhance productivity, health, and overall satisfaction.

1. Temperature and Thermal Comfort

Understanding Thermal Sensation

- Our bodies constantly seek thermal equilibrium.

- Designing spaces with optimal temperatures ensures comfort and focus.

Heating and Cooling Strategies

- Heating: Efficient heating systems prevent discomfort and cold stress.

- Cooling: Air conditioning, natural ventilation, and shading maintain comfort during hot weather.

2. Noise Control and Acoustics

Noise Pollution

- Excessive noise disrupts concentration and increases stress.

- Soundproofing materials, layout adjustments, and quiet zones mitigate noise.

Office Acoustics

- Sound Absorption: Carpets, curtains, and acoustic panels reduce echoes.

- Background Noise: White noise or calming music masks distracting sounds.

3. Air Quality and Ventilation

Indoor Air Quality (IAQ)

- Proper ventilation prevents stale air and pollutants.

- Regular maintenance of HVAC systems ensure clean air circulation.

Natural Ventilation

- Openable windows and cross-ventilation improve IAQ.

- Fresh air invigorates the mind and body.

4. Lighting and Circadian Rhythms

Natural Light Exposure

- Sunlight regulates our internal clock (circadian rhythm).

- Position workstations near windows to maximize daylight exposure.

Artificial Lighting

- Color Temperature: Cool white light for focus, warm light for relaxation.

- Task Lighting: Adjustable desk lamps for specific work areas.

5. Future Innovations

Biophilic Design

- Integrating nature into workspaces (plants, natural materials).

- Boosts mood, creativity, and cognitive performance.

Smart Buildings

- Sensors adjust lighting, temperature, and ventilation based on occupancy.

- Energy-efficient and user-friendly spaces.

Indoor air pollutants can significantly impact our health and well-being. Table 1 provides a quick reference to the potential health impacts of these common indoor air pollutants.

Table 1: Health impacts common indoor air pollutants

Indoor Air Pollutant	Description	Potential Health Impact
Asbestos	Found in older buildings, insulation, and construction materials.	Prolonged exposure can lead to lung diseases, including cancer.
Biological Pollutants	Includes mold, pollen, pet dander, and dust mites.	Allergens that can trigger respiratory issues and allergies.
Carbon Monoxide (CO)	Produced by incomplete combustion (e.g., gas stoves, heaters, fireplaces).	Odorless and highly toxic; exposure can be fatal.
Formaldehyde/Pressed Wood Products	Found in furniture, cabinets, and building materials.	Long-term exposure may cause respiratory problems and irritation.
Lead (Pb)	Common in older homes with lead-based paint or pipes.	Neurological and developmental effects, especially in children.
Nitrogen Dioxide (NO_2)	Emitted from gas stoves, car exhaust, and tobacco smoke.	Aggravates respiratory conditions like asthma.
Pesticides	Used for pest control but can linger in indoor air.	Health risks vary depending on the specific pesticide.
Radon (Rn)	Naturally occurring radioactive gas seeping from soil and rock.	Prolonged exposure increases the risk of lung cancer.
Indoor Particulate Matter	Includes fine particles ($PM_{2.5}$) from cooking, smoking, and outdoor pollution.	Can cause respiratory problems and cardiovascular effects.
Secondhand Smoke/Environmental Tobacco Smoke	Harmful chemicals from tobacco smoke.	Has been linked to lung cancer, heart disease, and respiratory issues.
Volatile Organic Compounds (VOCs)	Emitted from paints, cleaning products, and synthetic materials.	Can cause headaches, eye irritation, and long-term health effects.

It's important to maintain good indoor air quality to protect your health and well-being. Therefore, improving air quality in your home office is essential for your health and productivity. Here are some practical steps you can take:

Ventilation:

- Open Windows: Regularly open windows to allow fresh air circulation.
- Use Fans: Ceiling fans or portable fans help disperse indoor air pollutants.
- Air Purifiers: Consider using air purifiers with HEPA filters to remove particles and allergens from the air.

Control Sources of Pollution:

- Avoid Smoking: If you smoke, do it outside to prevent secondhand smoke indoors.
- Limit VOCs: Volatile organic compounds (VOCs) are emitted from paints, cleaning products, and furniture. Choose low-VOC or VOC-free options.
- Keep Clean: Regularly dust, vacuum, and mop to reduce dust mites and other allergens.

Humidity Control:

- Maintain Ideal Humidity: Aim for 30-50% relative humidity. Use a dehumidifier if it's too humid or a humidifier if it's too dry.
- Prevent Mold: Fix any leaks promptly and ensure good ventilation in bathrooms and kitchens.
- Plants: Indoor plants not only add greenery but also help improve air quality by absorbing pollutants. Consider plants like snake plants, peace lilies, and spider plants.

Natural Cleaning Alternatives:

- Use natural cleaning products like vinegar, baking soda, and lemon juice. Avoid harsh chemicals that release harmful fumes into the air.

Position Your Workspace Strategically:

- Place your desk near windows to benefit from natural light and fresh air. Avoid positioning your desk near potential sources of pollution (e.g., printers, copiers).

Keeping a healthy home office environment contributes to overall well-being and work performance.

CHAPTER EIGHT

Musculoskeletal Disorders and Injury Prevention

Safeguarding the Well-Being of Workers

In this chapter, we delve into the critical topic of musculoskeletal disorders (MSDs) and effective strategies for preventing workplace injuries. By understanding risk factors and implementing preventive measures, we create safer and healthier work environments.

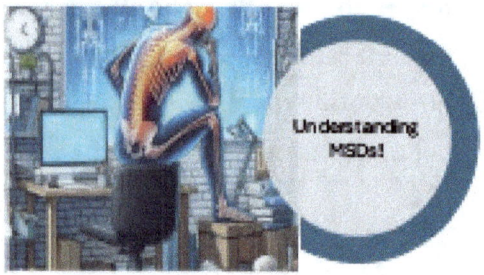

1. Understanding Musculoskeletal Disorders (MSDs)

What Are MSDs?

- MSDs affect muscles, tendons, ligaments, nerves, and joints.

- Common examples include back pain, carpal tunnel syndrome, and repetitive strain injuries.

Risk Factors

- Repetitive Movements: Prolonged repetitive tasks strain muscles and tendons.

- Awkward Postures: Poor ergonomics lead to unnatural body positions.

- Forceful Exertions: Lifting heavy objects without proper technique.

- Vibration Exposure: Prolonged use of vibrating tools.

2. Injury Prevention Strategies

Ergonomic Workstations

- Proper Seating: Adjustable chairs with lumbar support.

- Monitor Position: Eye-level placement to reduce neck strain.

- Keyboard and Mouse: Neutral wrist positions.

Lifting Techniques

- Bend Your Knees: Use leg muscles, not your back.

- Keep Loads Close: Minimize reaching and twisting.

- Team Lifting: Share heavy loads with colleagues.

Stretching and Exercise

- Dynamic Stretching: Warm up before work.

- Strengthen Core Muscles: Abdominals and back muscles support posture.

Workplace Design

- Anti-Fatigue Mats: Reduce strain from prolonged standing.

- Proper Lighting: Avoid eye strain and glare.

- Noise Reduction: Quieter environments reduce stress.

3. Early Intervention and Reporting

Recognize Symptoms

- Educate workers about early signs of MSDs.

- Encourage reporting discomfort promptly.

Prompt Treatment

- Seek medical attention if symptoms persist.

- Early intervention prevents chronic conditions.

4. Training and Education

Employee Training

- Train workers on proper lifting techniques.

- Educate about ergonomics and injury prevention.

Supervisor Awareness

- Supervisors play a crucial role in identifying risks.

- Regular safety assessments and communication are essential.

When sitting at desks, people often make several common posture mistakes that can lead to discomfort and health issues as illustrated in Table 2.

Table 2: Posture mistakes people often make when sitting at desks and their solutions.

Common Posture Mistakes	Potential Issues	Solutions
Crossing Your Legs	Extra stress on hips and lower back.	Sit with both feet flat on the floor. Use a footrest if needed.
Sitting Down for Too Long	Muscle tightness and discomfort.	Take breaks every 30 minutes to get up, stretch, and increase blood flow.
Hunching Over Your Desk	Strains your spine and neck, leading to pain.	Set your chair between 90 and 120 degrees so you can sit with a straight or slightly reclined back.
Leaning Your Head Forward	Can cause neck strain.	Position your monitor at eye level to maintain a neutral neck position.
Looking at a Poorly Positioned Monitor	Can strain your neck and eyes.	Position the top third of the screen at eye level.
Sitting on Non-Ergonomic Chairs	Can contribute to poor posture.	Invest in ergonomic chairs that support your entire body and allow adjustments for optimal comfort.

Consistent practice and awareness are key to maintaining good posture.

CHAPTER NINE

Ergonomics in Product Design: Creating User-Centric Solutions

Understanding Ergonomics

Ergonomics, also known as human factors engineering, focuses on designing products and systems that fit seamlessly into users' lives. By considering human capabilities, limitations, and interactions, product designers create functional, comfortable, and intuitive solutions. Let's explore how ergonomic principles enhance consumer products, tools, and gadgets.

1. User-Centered Design

Know Your Users

- Understand the target audience: their demographics, needs, and preferences.

- Conduct user research, surveys, and usability testing to gather insights.

Anthropometry and Biomechanics

- Anthropometric Data: Use body measurements to determine product dimensions (e.g., chair height, handle grip).

- Biomechanics: Consider how users interact physically with the product (e.g., lifting, typing).

2. Key Ergonomic Considerations

Form and Function

- Aesthetics: A visually appealing design enhances user satisfaction.

- Functionality: Prioritize usability and practicality.

Comfort and Safety

- Comfort: Ergonomic shapes, padding, and materials improve comfort.

- Safety: Rounded edges, non-slip surfaces, and proper weight distribution prevent accidents.

Accessibility

- Design for diverse abilities (e.g., consider users with mobility challenges).

- Ensure controls, buttons, and displays are easy to reach and operate.

3. Ergonomic Product Categories

Furniture and Seating

- Chairs: Lumbar support, adjustable height, and seat depth.

- Desks: Proper height, legroom, and cable management.

Handheld Devices

- Smartphones: Consider grip size, button placement, and screen readability.

- Tools: Ergonomic handles reduce strain during prolonged use.

Wearable Technology

- Fitness Trackers: Comfortable straps, accurate sensors, and intuitive interfaces.

- Smartwatches: Lightweight, customizable, and user-friendly.

4. Iterative Design and Prototyping

User Testing

- Prototype early and often.

- Gather feedback from real users to refine designs.

Iterate Based on Feedback

- Address pain points and usability issues.

- Continuously improve the product based on user insights.

5. Future Trends

Virtual Reality (VR) and Augmented Reality (AR)

- Designing comfortable headsets and controllers.

- Balancing aesthetics with functionality.

Sustainable Materials

- Prioritizing eco-friendly materials and recyclability.

- Minimizing environmental impact.

In general, it can be said that Ergonomics isn't just about aesthetics—it's about creating products that enhance users' lives. By integrating ergonomic principles into every stage of design, we build a future where technology seamlessly aligns with human needs. Let's explore some examples of poorly designed products from an ergonomic perspective in Table 3:

Table 3: Potential impacts caused by common ergonomic mistakes.

Poorly Designed Product	Issue	Potential Impact
Non-Adjustable Office Chairs	Chairs without adjustable features (height, lumbar support, arm rests) force users into uncomfortable positions.	Back pain, poor posture, and reduced productivity.
Small Computer Keyboards	Keyboards with cramped layouts or tiny keys strain users' wrists and fingers.	Increased risk of repetitive strain injuries (RSIs) like carpal tunnel syndrome.
Mobile Devices with Poor Screen Visibility	Smartphones or tablets with glare-prone screens or small fonts force users to squint or hunch over.	Eye strain, neck pain, and discomfort.
Heavy Power Tools with Poor Handles	Tools like drills or chainsaws with poorly designed handles cause excessive wrist and arm strain.	Fatigue, reduced control, and potential accidents.
Non-Ergonomic Computer Mouse Shapes	Mice that don't fit the natural curve of the hand lead to awkward wrist positions.	Wrist pain and discomfort during prolonged use.
Overly Complex Remote Controls	TV or home theater remotes with too many buttons confuse users.	Frustration, difficulty finding the right controls, and reduced usability.
Uncomfortable Car Seats	Car seats lacking proper lumbar support or adjustable features cause discomfort during long drives.	Back pain, fatigue, and reduced driving safety.
Poorly Designed Backpacks	Backpacks with thin straps or inadequate padding strain shoulders and back.	Shoulder pain, poor weight distribution, and potential spinal issues.

Essentially ergonomic design prioritizes user comfort, safety, and usability. Avoiding the mistakes highlighted ensures better product experiences for everyone. In Table 4 examples and benefits of well-designed ergonomic products.

Table 4: Features of well-designed ergonomic products that prioritize user comfort, safety, and usability

Ergonomic Products	Features	Benefits
Ergonomic Chairs	*Smugdesk Ergonomic Chair*: Lumbar support, memory foam seat, adjustable headrest, mesh construction. *Duramont Ergonomic Adjustable Office Chair*: Extensive adjustability options, lumbar support, breathable mesh design.	Ensures comfort during long hours of work, highly adaptable, promotes good posture.
Ergonomic Keyboards and Mice	*Split Keyboards*: Allows a more natural hand position. *Ergonomic Mouse Designs*: Contoured mice with thumb rests and customizable buttons.	Reduces wrist strain, enhances comfort.
Standing Desks and Converters	*Height-Adjustable Desks*: Allows users to switch between sitting and standing positions. *Monitor Stands*: Elevates monitors to eye level.	Promotes better posture, reduces sedentary behavior, prevents neck strain.
Wearable Fitness Trackers and Smartwatches	*Comfortable Straps*: Soft, adjustable straps. *Intuitive Interfaces*: User-friendly touchscreens and customizable features.	Ensures all-day wearability, easy to use.
Ergonomic Backpacks	*Padded Straps and Back Panels*: Distributes weight evenly. *Multiple Compartments*: Organizes belongings.	Reduces pressure on shoulders and spine, prevents strain from uneven loads.
Ergonomic Kitchen Tools	*Ergonomic Knives*: Handles designed for comfortable grip. *Non-Slip Cutting Boards*: Prevents accidents during food preparation.	Reduces wrist strain, enhances safety.
Office Accessories	*Footrests*: Elevates feet. *Document Holders*: Keeps documents at eye level.	Improves circulation, reduces pressure on the lower back, prevents neck strain.

As established, ergonomic design prioritizes user needs and well-being. The examples in Table 4 demonstrate how thoughtful design enhances our daily experiences. Now, how does one ensure the products meet user needs and well-being? This is done conducting evaluation!

Effective methods for evaluating ergonomic products:

1. User Testing and Feedback:

Observation: Observe users interacting with the product in real-world scenarios.

Surveys and Interviews: Collect feedback on comfort, usability, and any pain points.

Iterate Based on User Insights: Use this feedback to refine the design and address any issues.

2. Anthropometric Measurements:

Body Dimensions: Compare product dimensions to standard anthropometric data (e.g., seat height, handle grip).

Postural Deviation: Assess how well the product supports natural body postures.

3. Biomechanical Analysis:

Muscle Activation: Measure muscle activity during product use (e.g., EMG data).

Joint Positions: Evaluate joint angles and movements.

Force Distributions: Assess force distribution (e.g., hand grip, pressure over the palm).

4. Vibration Attenuation Assessment:

For tools or equipment with vibration (e.g., hammers, power tools), measure vibration transmitted to the hand and upper extremity.

Compare different product designs for vibration reduction.

5. Comparative Studies:

Benchmarking: Compare the product's ergonomic features against competitors' products.

Safety Standards: Evaluate whether the product meets published safety standards.

6. Subjective Assessment:

User Perception: Gather subjective rankings from users.

Comfort and Satisfaction: Assess how comfortable and user-friendly the product feels.

When conducting such evaluation, one needs to *avoid the following common mistakes* that can compromise user comfort and safety:

1. Focusing Solely on Aesthetics:

Prioritizing visual appeal over functionality can lead to uncomfortable or inefficient designs.

Solution: Balance aesthetics with usability and ergonomic features.

2. Ignoring User Feedback:

Neglecting user insights can result in products that don't meet real-world needs.

Solution: Regularly gather feedback through testing, surveys, and observations.

3. Overlooking Adjustability:

Non-adjustable products limit customization for different users.

Solution: Design products with adjustable features (e.g., chair height, monitor tilt).

4. Neglecting Safety Standards:

Ignoring safety guidelines can lead to hazardous designs.

Solution: Ensure products meet necessary ergonomic and safety standards.

5. Not Considering Diverse Abilities:

Designing for a narrow range of users excludes those with different abilities.

Solution: Create inclusive designs that accommodate various physical capabilities.

CHAPTER TEN

Ergonomics in Healthcare and Special Populations

Designing for Comfort, Safety, and Inclusivity.

In this chapter, we explore the critical role of ergonomics in healthcare settings, with a specific focus on aging populations and individuals with disabilities. By prioritizing user-centered design, we create environments and products that enhance well-being, promote independence, and ensure equitable access to care.

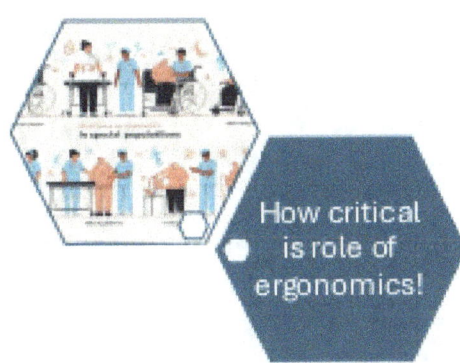

1. Ergonomics in Healthcare Settings

Hospital Environments

- Patient Beds: Design adjustable beds with easy-to-use controls for patients and caregivers.

- Wheelchair Accessibility: Ensure corridors, doorways, and restrooms accommodate wheelchairs.

- Lighting and Noise Control: Optimize lighting and minimize noise to aid patient recovery.

Medical Equipment

- Diagnostic Machines: Position screens, buttons, and handles for ease of use.

- IV Poles and Infusion Pumps: Design stable, maneuverable equipment for healthcare professionals.

2. Aging Populations

Mobility and Fall Prevention

- Grab Bars: Install sturdy grab bars in bathrooms and hallways.

- Non-Slip Flooring: Choose slip-resistant materials to prevent falls.

- 43 -

Assistive Devices

- Canes and Walkers: Design lightweight, adjustable devices with comfortable grips.

- Hearing Aids and Glasses: Prioritize user-friendly controls and comfort.

3. Individuals with Disabilities

Accessible Workstations

- Adaptive Keyboards and Mice: Consider users with limited dexterity.

- Screen Readers: Ensure digital interfaces are compatible with screen-reading software.

Customizable Interfaces

- Font Size and Contrast: Allow users to adjust text size and color contrast.

- Voice Commands: Incorporate voice-activated controls for hands-free interaction.

4. Inclusive Design Principles

Universal Design

- Equitable Use: Products should be usable by all, regardless of ability.

- Flexibility in Use: Accommodate diverse needs and preferences.

Collaboration with Users

- Co-Design: Involve end-users in the design process.

- Feedback Loops: Continuously gather insights and iterate based on user experiences.

5. Future Challenges and Innovations

Virtual Reality (VR) for Rehabilitation

- VR-based exercises for physical therapy and cognitive training.

- Customizable experiences for various conditions.

Smart Homes and Assistive Technologies

- Voice-controlled home automation for individuals with mobility challenges.

- Wearable health monitors for real-time data collection.

Practical Tips and Examples

Ergonomics in healthcare extends beyond physical comfort—it's about dignity, independence, and quality of life. By embracing inclusive design, we create a world where everyone receives the care they deserve. A practical example can be designing an accessible waiting room for a healthcare facility is crucial to ensure comfort, safety, and inclusivity for all visitors. Here are some key considerations and design tips:

1. Wayfinding and Clear Signage:

Intuitive Layout: Arrange seating areas, reception desks, and restrooms in a logical flow.

Clear Signage: Use large, easy-to-read signs with high contrast to guide visitors.

2. Screening Areas:

Permanent Screening Stations: Include temperature check and health assessment stations near the entrance.

Privacy and Comfort: Design these areas to comply with privacy laws (such as HIPAA) and create a welcoming atmosphere.

3. Accessible Amenities:

Mobility-Friendly Design: Ensure unobstructed pathways, self-opening double doors, and ample space for wheelchairs and strollers.

Clear Line of Sight: Position reception desks within sight of the entry to assist visitors promptly.

Proximity to Restrooms: Place restrooms near waiting areas to minimize wandering.

4. Comfortable and Attractive Seating:

Diverse Seating Options: Offer a mix of chairs, benches, and sofas to accommodate different preferences.

Comfortable Cushions: Provide well-padded seats for extended waiting times.

Weight Capacity: Ensure chairs can support various body sizes.

5. Lighting and Acoustics:

Natural Light: Maximize daylight with large windows or skylights.

Noise Control: Use acoustic materials to reduce echoes and create a calm environment.

6. Inclusive Design for Special Populations:

Aging Populations: Consider higher seat height and stable armrests for ease of getting up.

Individuals with Disabilities: Prioritize clear pathways, accessible seating, and adjustable tables.

7. Art and Décor:

Welcoming Atmosphere: Use soothing colors, artwork, and greenery to create a pleasant ambiance.

Local Touches: Incorporate regional or cultural elements to make visitors feel at home.

An accessible waiting room not only serves patients but also reflects the healthcare facility's commitment to providing quality care for everyone. *When designing waiting rooms, innovative seating solutions can enhance comfort, aesthetics, and functionality. Here are some creative ideas to consider:*

1. Clustered Seating Areas:

Instead of traditional rows of chairs, create organic clusters of seats. These groupings allow families to gather and discuss treatment options without worrying about privacy.

2. Zero Gravity Chairs:

These chairs provide exceptional comfort by distributing body weight evenly and reducing pressure on the spine. They are especially beneficial for patients with back pain or mobility issues.

3. Modular Seating:

Use modular furniture that can be rearranged easily. This versatility allows you to adapt the seating arrangement based on the number of visitors or specific needs.

4. Charging Stations with Seating:

Combine seating with charging stations. Provide comfortable chairs or benches near power outlets so visitors can charge their devices while waiting.

5. Assistive Seating for Special Populations:

Bariatric Chairs: Accommodate larger individuals comfortably.

Child-Friendly Seating: Include child-sized chairs or play areas for pediatric waiting rooms.

Wheelchair-Accessible Seating: Ensure there's enough space for wheelchair users and their companions.

6. Multi-Functional Furniture:

Storage Benches: Combine seating with storage compartments for bags, coats, or personal items.

Flip-Up Seats: Install seats that can be flipped up to create additional space when not in use.

7. Biophilic Seating Design:

Living Walls: Incorporate vertical gardens or greenery into seating areas.

Natural Materials: Use wood, bamboo, or other natural materials for a calming effect.

Innovative seating solutions not only improve the waiting experience but also contribute to the overall ambiance of the healthcare facility. They can be both functional and delightful, providing visitors with a positive experience while they wait. *Are you considering outdoor spaces!?* Here are some innovative ideas:

1. Outdoor Courtyards and Gardens:

Green Spaces: Design lush gardens with seating areas, shade, and calming plants.

Water Features: Incorporate fountains, ponds, or small streams for a soothing atmosphere.

2. Interactive Art Installations:

Sculptures: Place unique sculptures or kinetic art pieces that engage visitors.

Interactive Murals: Allow visitors to contribute to a communal mural or chalkboard wall.

3. Mobile Seating Pods:

Moveable Benches: Install benches on wheels that visitors can rearrange as needed.

Swing Seats: Hang swing-like seats for a playful touch.

4. Outdoor Lounges:

Comfortable Sofas: Arrange outdoor sofas with cushions and throw pillows.

Fire Pits or Heaters: Extend usability during colder months.

5. Shaded Pavilions or Canopies:

Retractable Canopies: Provide shade and protection from rain or sun.

Solar-Powered Charging Stations: Combine seating with charging capabilities.

6. Outdoor Reading Nooks:

Bookshelves or Cubbies: Create small nooks with books or magazines.

Hammocks or Hanging Chairs: Encourage relaxation while waiting.

7. Themed Waiting Areas:

Beach-Inspired: Use sand, beach chairs, and umbrellas for a coastal vibe.

Urban Oasis: Incorporate cityscape murals, streetlamps, and café-style seating.

CHAPTER ELEVEN

Enhancing Ergonomics with Exoskeletons

Introduction

Exoskeletons, as wearable devices that augment human performance, have a profound impact on ergonomics, the science of designing and arranging things people use so that the people and things interact most efficiently and safely.

Exoskeletons for Ergonomic Enhancement

Exoskeletons can enhance ergonomics in several ways:

- Reducing Physical Strain: By providing physical support and augmenting strength, exoskeletons can reduce the strain on the user's body, preventing injuries and reducing fatigue.

- Improving Posture: Exoskeletons can help maintain proper posture, reducing the risk of musculoskeletal disorders.

- Increasing Efficiency: By augmenting human strength and endurance, exoskeletons can increase efficiency and productivity in tasks that require physical exertion.

Applications of Exoskeletons in Ergonomics

Exoskeletons have found applications in various fields where ergonomics is crucial:

- Industrial Settings: Exoskeletons are used to assist workers in lifting heavy objects, reducing the risk of injury, and improving productivity.

- Healthcare: In rehabilitation, exoskeletons can provide support and assistive force to help patients regain their mobility.

- Military: Exoskeletons can enhance soldiers' physical capabilities, reducing fatigue and increasing endurance.

Challenges and Future Directions

While exoskeletons hold great promise for enhancing ergonomics, there are challenges to be addressed:

- User Comfort: Ensuring that the exoskeleton is comfortable and does not restrict movement is crucial for user acceptance.

- Cost and Accessibility: Prohibitive costs can be a barrier to widespread adoption of exoskeleton technology.

- Regulation and Standards: There is a need for regulatory standards to ensure the safety and effectiveness of exoskeletons.

Exoskeletons offer a promising avenue for enhancing ergonomics in various fields. As technology advances, we can expect to see more ergonomic applications of exoskeletons in the future.

CHAPTER TWELVE

Leveraging AI for Ergonomics

Introduction

Artificial Intelligence (AI) has the potential to revolutionize the field of ergonomics, making workplaces safer, more efficient, and more comfortable. This chapter explores how AI can be leveraged to enhance ergonomics.

AI in Ergonomic Assessment

AI can be used to assess ergonomic risks in real-time:

- Posture Analysis: AI algorithms can analyze video feeds to identify poor postures that could lead to musculoskeletal disorders.

- Workplace Design: AI can help design workspaces that are more ergonomic, reducing the risk of injury and improving productivity.

AI in Ergonomic Training

AI can also play a role in ergonomic training:

- Personalized Training: AI can tailor ergonomic training to individual needs, improving effectiveness.

- Real-time Feedback: AI can provide real-time feedback during training, helping individuals correct their posture and technique.

AI in Ergonomic Product Design

AI can contribute to the design of ergonomic products:

- Predictive Modeling: AI can predict how different designs will impact ergonomics, helping designers make informed decisions.

- User Testing: AI can analyze user testing data to identify potential ergonomic issues.

Challenges and Future Directions

While AI holds great promise for enhancing ergonomics, there are challenges to be addressed:

- Data Privacy: Protecting the privacy of individuals is crucial when using AI for ergonomic assessment and training.

- Algorithm Bias: Ensuring that AI algorithms do not unfairly disadvantage certain groups of people is a key challenge.

Practical Tips and Application:

AI offers exciting possibilities for enhancing ergonomics. As technology advances, we can expect to see more applications of AI in this field in the future. Examples of AI applications in ergonomics:

1. Automated Posture Analysis: AI-powered systems can analyze video footage of workers in real-time to identify risky postures and movements that may lead to musculoskeletal disorders.
2. Healthcare AI: AI has been used to develop and adopt artificial intelligence technologies in healthcare. This includes imaging and diagnostics, prehospital triage, care management, and mental health.
3. Fighter Aircraft Displays: AI has been applied to automate systems in fighter aircraft displays to improve pilot safety by controlling the information provided to the pilot depending on the circumstances.
4. Workplace Safety: Computer vision technology has made substantial strides in enhancing ergonomic practices by integrating advanced AI capabilities. This technology acts as a 24/7 ergonomist, detecting potential postural and physical hazards, and fostering healthier work environments through proactive intervention.
5. Human-Centered Work: AI technologies critically deal with human-centered, effective as well as efficient work.

These examples illustrate how AI can be leveraged to enhance ergonomics, making workplaces safer, more efficient, and more comfortable. It is important to understand *challenges present when Implementing AI for ergonomics* and strategies for mitigation. Some of these challenges include:

1. Data Privacy: Protecting the privacy of individuals is crucial when using AI for ergonomic assessment and training.

2. Algorithm Bias: Ensuring that AI algorithms do not unfairly disadvantage certain groups of people is a key challenge.

3. Black Box Nature of AI: The opaque nature of AI algorithms can lead to distrust and uncertainty among users.

4. Ethical Concerns: Accountability for AI-generated decisions, and potential biases in training data raise important considerations.

5. Weak Evidence Base: The evidence base for the effectiveness of deep learning algorithms remains weak and is at high risk of bias, because there are few independent prospective evaluations.

6. Real-world Assessment: The performance, usability, and safety of these technologies can only be reliably assessed in real-world settings, where teams of healthcare workers and AI technologies co-operate and collaborate to provide a meaningful service.

7. Human Factors: Addressing challenges related to human factors, safety, and user-centered design is crucial when integrating AI tools with ergonomics.

These challenges need to be addressed to fully leverage the potential of AI in enhancing ergonomics. Here are some strategies to mitigate privacy concerns:

1. Develop a Comprehensive AI Use Policy: Draft a clear policy that governs the use of AI within the organization.

2. Conduct Privacy Impact Assessments (PIAs): These assessments can help identify and mitigate privacy risks associated with AI.

3. Ensure Transparency and Consent: Users should be informed about how their data is being used and consent should be obtained.

4. Implement Robust Data Security Measures: Data security measures such as encryption can protect sensitive data.

5. Stay Updated on Regulations and Standards: Keeping up to date with the latest privacy regulations and standards can help ensure compliance.

6. Foster a Culture of Privacy: Encourage a culture within the organization that values and respects privacy.

7. Embed Privacy in AI Design: Integrate privacy considerations at the initial stages of AI system development.

8. Advanced Measures: Implement advanced measures such as face blurring, background blurring, and encryption ensures the protection of sensitive data collected through AI-driven ergonomics.

9. Clear Protocols and Opportunities for Collaboration and Feedback: By providing ongoing upskilling, issuing clear protocols, creating opportunities for collaboration and feedback, and focusing on transparency, companies can help ensure this technology results in more productivity and less harm.

CHAPTER THIRTEEN

Ergonomics Training

Introduction

Ergonomics training is a crucial aspect of creating a safe and efficient work environment. It involves educating employees about the principles of ergonomics and how to apply them in their daily tasks.

Importance of Ergonomics Training

Ergonomics training can lead to several benefits:

- Reduced Risk of Injury: By teaching employees how to work safely, ergonomics training can reduce the risk of work-related injuries.

- Increased Productivity: Employees who are trained in ergonomics are likely to be more efficient and productive.

- Improved Employee Well-being: Ergonomics training can lead to improved comfort and well-being, leading to higher job satisfaction.

Components of Ergonomics Training

Effective ergonomics training should include the following components:

- Understanding Ergonomics: The training should start with a basic understanding of ergonomics and its importance.

- Identifying Ergonomic Risks: Employees should be trained to identify ergonomic risks in their work environment.

- Implementing Ergonomic Solutions: The training should include practical exercises on how to implement ergonomic solutions.

- Maintaining Ergonomic Practices: The training should emphasize the importance of maintaining ergonomic practices over time.

Challenges in Ergonomics Training

Despite its importance, there are several challenges in implementing ergonomics training:

- Lack of Awareness: Many employees may not be aware of the importance of ergonomics, making it difficult to engage them in training.

- Resource Constraints: Organizations may lack the resources needed to provide comprehensive ergonomics training.

- Sustainability: It can be challenging to ensure that employees continue to apply ergonomic principles after the training is over.

Despite these challenges, ergonomics training is a worthwhile investment. By educating employees about the principles of ergonomics and how to apply them, organizations can create a safer, more efficient, and more comfortable work environment. Innovative approaches to delivering ergonomics training are continually evolving with advancements in technology. Here are some examples:

1. Online Training: *With the rise of digital platforms, online training has become a popular* method for delivering ergonomics training. It offers flexibility, allowing employees to learn at their own pace and in their own time.

2. Virtual and Augmented Reality: These technologies provide immersive training experiences. They can simulate various work environments and ergonomic challenges, offering hands-on training in a controlled, risk-free setting.

3. AI and Machine Learning: AI can personalize training based on an individual's needs and progress. Machine learning algorithms can analyze data from various sources (like wearable sensors) to provide insights and recommendations.

4. Blended Learning: This approach *combines online self-paced learning with live instructor-led training*. It provides the flexibility of online training with the direct interaction and feedback of traditional classroom training.

5. Exoskeletons and Wearable Technology: These technologies can provide real-time feedback on an individual's posture and movement, allowing for immediate correction and learning.

6. Computer Vision: This technology can assess hazards and recommend ergonomic improvements, providing practical, real-world training.

Remember, the choice of method often depends on an organization's specific needs and constraints. Here are some examples (as of time of writing this book) of blended learning applied in the field of ergonomics:

1. University of Nottingham's Distance Learning Courses: The University of Nottingham has been offering distance learning courses in Human Factors and Ergonomics since 2001. These courses, recognized by the Chartered Institute of Ergonomics & Human Factors, are aimed primarily at practitioners. They offer a flexible route to gaining formal qualifications.

2. ClassPoint's Blended Learning Models: ClassPoint provides specific blended learning examples, showcasing how their tool seamlessly integrates into these activities. They cover various models like flipped classrooms, station rotation, and collaborative projects.

3. Workplace Blended Learning Models: There are different models of blended learning that have been successfully implemented in the workplace. These include the flipped classroom model, station rotation model, lab rotation model, flex model, online lab model, and individual rotation model.

4. Alternative Energy Sources Unit: A teacher created a digital playlist of videos for students to work through to complete a unit on alternative energy sources, then take an exam on the material in class.

CHAPTER FOURTEEN

Ergonomics and Remote Work

Adapting to New Environments: Strategies for Setting Up an Ergonomic Home Office

The shift to remote work has brought the importance of a well-designed home office into sharp focus. One should explore how to create an ergonomic workspace that mimics the benefits of a traditional office setting. Ensuring correct selection and arrangement of furniture, the importance of natural lighting, and follow tips for creating a distraction-free zone.

Technology and Equipment: Choosing the Right Tools to Maintain Ergonomics in a Remote Setting

With the right technology and equipment, maintaining an ergonomic workspace is achievable. Ensure all the essential tools for remote work, from ergonomic chairs and adjustable desks to keyboards and mice designed for comfort are present. Consider the role of technology in staying connected and productive.

Physical and Mental Well-being: Balancing Work and Life to Prevent Burnout and Maintain Physical Health

Remote work can blur the lines between personal and professional life, leading to burnout and health issues. Establish a work-life balance, incorporating regular breaks, and staying physically active. Emphasis be on the need for mental health breaks and setting boundaries.

Virtual Collaboration: Ergonomic Considerations for Long-Duration Virtual Meetings and Collaboration

Long hours of virtual meetings can be taxing. Ensure setting up a conducive environment for virtual collaboration, including camera positioning, proper lighting, and maintaining good posture. Take short breaks during prolonged sessions to reduce strain.

Self-Assessment and Adjustment: Techniques for Individuals to Self-Assess Their Ergonomic Setup and Make Necessary Adjustments

Individuals should be empowered to take charge of their ergonomic well-being. Conducting simple self-assessment techniques to evaluate one's home office setup which will provide guidance on making incremental adjustments to improve comfort and productivity should be prioritized.

This chapter complements what has already been learnt (existing knowledge) on office ergonomics. It equips remote workers with the knowledge to create a healthy and sustainable work environment, ensuring that they can thrive in the era of remote work.

CHAPTER FIFTEEN:

Answering Ergonomic Questions

Introduction

Answering ergonomic questions requires a deep understanding of the principles of ergonomics and its application in various contexts. This chapter provides a guide on how to effectively answer ergonomic questions.

Understanding the Question

The first step in answering an ergonomic question is to understand what is being asked. This involves identifying the key elements of the question and the context in which it is being asked.

Gathering Information

Once the question is understood, the next step is to gather relevant information. This could involve:

- Research: Look up reliable sources to find information related to the question.

- Observation: In some cases, direct observation of the ergonomic issue may be necessary.

- Consultation: Consult with experts or refer to ergonomic guidelines and standards.

Formulating the Answer

After gathering the necessary information, formulate a clear and concise answer. The answer should directly address the question and provide practical recommendations when applicable.

Verifying the Answer

Before providing the answer, verify its accuracy. This could involve cross-checking with multiple sources or seeking feedback from experts.

Delivering the Answer

Finally, deliver the answer in a clear and understandable manner. Use simple language and avoid jargon as much as possible. If the answer involves complex concepts, consider using diagrams or examples to aid understanding.

Answering ergonomic questions is a skill that requires knowledge, research, and clear communication. By following these steps, you can provide accurate and helpful answers to ergonomic questions. Use the tips learnt to answer the following questions on ergonomics in relation to overall improvement:

1. How does the implementation of ergonomics in the workplace contribute to overall productivity?

2. What are some examples of ergonomic improvements that have significantly reduced work-related injuries?

3. How can ergonomics be integrated into the design process of products to improve user experience and satisfaction?

4. What role does ergonomics play in the development and application of new technologies such as AI and VR?

5. How can training in ergonomics lead to long-term improvements in employee health and well-being?

6. How can workplace design reduce physical strain?

7. What adjustments can be made to improve posture while sitting at a desk?

8. Are there ergonomic solutions to alleviate repetitive strain injuries?

9. How does lighting affect work comfort and efficiency?

10. What are the best practices for setting up a home office ergonomically? How can screen placement reduce eye strain?

11. What are effective stretches for someone who works at a computer all day?

12. Can ergonomic tools improve hand and wrist health?

13. How does chair design influence back health?

14. What is the impact of monitor height on neck position?

15. How can footrests contribute to better seating ergonomics?

16. What role does keyboard design play in preventing carpal tunnel syndrome?

17. How can noise levels be managed to improve concentration and comfort?

18. What are the ergonomic considerations for a standing desk?

19. How can the layout of a workspace promote movement and reduce sedentary behavior?

20. What factors should be considered when selecting ergonomic office equipment?

21. How can the arrangement of tools and materials minimize unnecessary motion?

22. What are the guidelines for optimal desk space to prevent clutter and overreach?

23. How can workplace ergonomics be tailored to individuals with specific needs?

24. What strategies can be implemented to encourage regular breaks and reduce fatigue?

CHAPTER SIXTEEN

Self-Study Questions for each Chapter

Chapter 1: Revision exercise questions based on this chapter to help in *understanding and reviewing the key concepts* of the chapter: *"Introduction to Ergonomics."*

1. Define the term "ergonomics" and explain its Greek roots.

2. Discuss the historical context of ergonomics. How did the Industrial Revolution influence the formal study of ergonomics?

3. Who was Frederick Winslow Taylor and what contributions did he make to the field of ergonomics?

4. Explain the concept of "Human-Centered Design" in ergonomics.

5. What does a "good fit" mean in the context of ergonomic design? Provide examples.

6. How does ergonomics contribute to safety and injury prevention in the workplace?

7. Discuss the role of ergonomics in enhancing productivity and efficiency.

8. Why is ergonomics considered a multidisciplinary field? Which disciplines does it draw knowledge from?

9. How does ergonomics impact health and well-being?

10. Explain how ergonomics can lead to cost savings for organizations.

11. Why is ergonomics considered a necessity and not a luxury in today's work environments?

12. How can the principles of ergonomics be applied as our work environments evolve?

Chapter 2: Exercise questions based on the chapter *"Human Factors and Human-Machine Interaction"*: These questions are intended to deepen your understanding of human factors and human-machine interaction by *applying the concepts to practical scenarios*.

1. Usability Evaluation: Design a study to evaluate the usability of a new smartphone interface. What methods would you use to assess its intuitiveness and efficiency?

2. Safety Analysis: Analyze a case where poor system design led to an accident. How could human factors principles have prevented it?

3. UX Assessment: Choose a technology product and describe how its user experience could be improved. Consider aspects like visual design, feedback, and user satisfaction.

4. Visual Interaction: How do visual cues on a dashboard help in reducing cognitive load for drivers? Provide examples.

5. Auditory Interaction: Discuss the role of auditory signals in user interfaces. How can they enhance or hinder user interaction?

6. Physical Interaction: Compare the effectiveness of touchscreens versus physical buttons in a high-stress environment like an emergency room.

7. Teaming Dynamics: Describe a scenario where teaming between humans and machines leads to improved decision-making. What factors contribute to this synergy?

8. Human Performance Enhancement: Propose a technology that could augment human capabilities in a specific profession. How would it improve productivity and accuracy?

9. Shared Control: Debate the pros and cons of shared control in autonomous vehicles. How does it impact safety and user empowerment?

10. Designing for Diversity: How can designers ensure that a new virtual reality headset accommodates a wide range of user sizes and preferences?

11. Error Management: Discuss how an understanding of human error can inform the design of a medical device to reduce the likelihood of operator mistakes.

12. Cognitive Load: Analyze how the design of a complex software application can be improved to reduce cognitive load for new users.

13. Adaptive Systems: Propose a feature for a car's navigation system that adapts to the driver's habits and preferences. How would this feature learn and evolve over time?

14. Clinical HMI Applications: How can human-machine interaction principles be applied to improve the usability of prosthetic devices for amputees?

15. Personalized Interaction: Design a personalized user interface for an online learning platform. What factors would you consider tailoring the experience to individual learning styles?

16. Model-Based Interface Design: Explain the benefits of using formal models in the design of an aircraft's cockpit interface.

17. Trust in Automation: What strategies can be employed to build trust in automated financial advising systems among users?

18. Physical Ergonomics: How does the physical design of a game controller affect the user's comfort and performance during extended periods of use?

19. Human-Agent Teaming: Describe how a human-agent teaming approach can enhance the efficiency of a warehouse logistics system.

20. Human-Agent Interaction: Evaluate the trustworthiness of a virtual assistant. What elements affect user trust and effective communication?

Chapter 3: *"Anthropometry: Designing for Human Proportions"*: The questions are designed to help learners *apply the concepts learned in the chapter to real-world scenarios* and deepen their understanding of anthropometric principles in design.

1. Historical Significance: How did ancient civilizations utilize anthropometry in their designs, and what impact did Leonardo da Vinci's sketches have on the field?

2. Understanding Body Dimensions: List and describe five different body dimensions that are crucial for designing ergonomic products and workspaces.

3. Proportional Design: Explain the role of the Golden Ratio and body segment ratios in ergonomic design. How do they influence aesthetics and functionality?

4. Ergonomic Workspace: What are the key anthropometric considerations when designing a desk and chair for an office environment?

5. Product Design for Diversity: How do anthropometric databases contribute to the design of universally comfortable products?

6. Vehicle Comfort: Discuss how anthropometric studies are used to optimize comfort for drivers and passengers in vehicle design.

7. Architecture and Public Spaces: Describe how anthropometry is applied in the design of doorways, stair risers, and room dimensions to accommodate diverse populations.

8. Addressing Population Diversity: Why is it important for designers to consider anthropometric data variation across different demographics?

9. Dynamic Anthropometry: How can adjustable furniture and adaptable spaces address the changing anthropometric needs of individuals?

10. Digital Anthropometry and Personalization: How has 3D scanning and virtual modeling changed the landscape of personalized design?

11. Inclusive Design: What is inclusive design, and how does it benefit people with disabilities or unique requirements?

12. Practical Application: If you were to design an ergonomic chair, which anthropometric measurements would be most relevant, and why?

13. Design Choices: Compare the advantages and disadvantages of designing for the average, adjustability, and extremes.

14. Future Trends: What future trends in anthropometry do you foresee impacting the design industry, and how?

15. Case Study Analysis: Analyze a case study where anthropometric data was crucial in solving a design problem.

Chapter 4: *"Biomechanics: Understanding Human Movement and Forces"*: Questions designed to help you reflect on the key concepts of biomechanics and apply them to practical scenarios.

1. Forces and Motion: How do external forces like gravity and friction affect human movement? Provide an example of how muscle tension influences body mechanics.

2. Newton's Laws: Explain how each of Newton's laws of motion applies to human movement. Can you give a practical example of one of the laws in action during a sporting activity?

3. Kinematics vs. Kinetics: What is the difference between kinematics and kinetics in the context of biomechanics? How do they complement each other in understanding human motion?

4. Gait Analysis: Describe the biomechanical factors that contribute to the efficiency of walking and running. How do stride length and foot strike affect locomotion?

5. Lifting Techniques: Why is it important to use proper lifting techniques? Discuss how biomechanical analysis can guide safe lifting practices.

6. Sports Biomechanics: Choose a sport and discuss how athletes can optimize their movements based on biomechanical principles.

7. Ergonomic Design: How does biomechanics influence the design of ergonomic chairs and desks? What are the benefits of aligning these items with natural body movements?

8. Manual Material Handling: What biomechanical principles should be considered when developing techniques for lifting, pushing, and pulling in the workplace?

9. Injury Mechanisms: How does biomechanics help in identifying risk factors for common injuries such as ACL tears or back pain?

10. Prosthetics Design: Explain how biomechanical principles are applied in the design of prosthetics and orthotic devices.

11. Virtual Reality: How is biomechanics used in virtual reality to train athletes or rehabilitate patients?

12. Bioinspired Design: Provide an example of bioinspired design that utilizes biomechanical principles. How does nature influence innovative materials and devices?

13. Biomechanics in Medicine: Discuss how biomechanics is used in the design of cell-based therapy delivery systems or drug delivery devices.

14. Predictive Modeling: How do computational biomechanics contribute to the predictive modeling of health and disease?

15. 3D Printing in Medicine: What role does biomechanics play in the design of 3D-printed biocompatible scaffolds for tissue engineering?

Chapter 5: Self-help questions to guide you through the process of creating an ergonomic workstation based on the chapter *"Workplace Design and Layout: Creating Ergonomic Workstations"*:

1. Ergonomic Assessment: Have I adjusted my chair to support my lower back and allow my feet to rest flat on the floor or on a footrest?

2. Monitor Positioning: Is my monitor placed at eye level and about 20 inches away to prevent neck strain and eye fatigue?

3. Workspace Organization: Have I arranged frequently used items within easy reach to minimize unnecessary movements and reduce the risk of strain?

4. Keyboard and Mouse Ergonomics: Is my keyboard slightly tilted away and my mouse positioned close to avoid overextending my wrists?

5. Lighting Considerations: Have I positioned my workstation to make the best use of natural light while ensuring that task lighting is available where needed?

6. Color Influence: Does the color scheme of my workspace align with the type of work I do, promoting either focus or creativity?

7. Noise Management: What steps can I take to reduce noise distractions in my workspace? Have I considered using acoustic panels or noise-canceling headphones?

8. Privacy Measures: If privacy is important in my work, have I created individual work zones or designated quiet rooms for concentration?

9. Practical Lighting Tips: How can I improve the lighting in my workspace to avoid glare and shadows, and what type of task lighting should I use for different workstations?

10. Continuous Improvement: What changes can I make right now to enhance the ergonomics of my workspace, and how can I continue to make improvements over time?.

Chapter 6: Self-study questions for the chapter *"Cognitive Ergonomics: Enhancing Mental Efficiency in Work Environments"*. Designed to encourage critical thinking and application of cognitive ergonomics principles to practical workplace scenarios.

1. Information Processing: How does cognitive ergonomics apply to the design of work environments with respect to information processing stages?

2. Attention and Focus: What workspace design elements should be considered to enhance focus and minimize distractions?

3. Heuristics and Biases: How can an understanding of heuristics and biases improve decision-making processes in the workplace?

4. Decision Support Systems: In what ways can technology like AI and data analytics be integrated to support decision-making and cognitive efficiency?

5. Cognitive Load: What strategies can be employed to balance mental resources and reduce unnecessary cognitive load during tasks?

6. Stress Management: How do high-stress environments impact cognitive function, and what techniques can mitigate this stress?

7. User Interface Design: Why is intuitive interface design crucial in minimizing cognitive effort, and how does it contribute to HCI?

8. Adaptive Systems: How do AI-driven systems adapt to user behavior, and in what ways do they optimize cognitive load?

9. Neuro-ergonomics: What is brain-computer interfaces (BCIs), and how might they change the way we interact with technology?

10. Collaborative Cognition: How does the study of group dynamics and distributed cognition contribute to our understanding of team information processing?

11. Building Psychological Safety: What are the benefits of creating a psychologically safe work environment, and how can it be achieved?

12. Effective Delegation: How does effective delegation contribute to team productivity and satisfaction?

13. Decision-Making Framework: What are the different decision-making frameworks, and how do they impact team dynamics?

14. Communication for Collaboration: How does open and transparent communication facilitate collaboration in the workplace?

15. Professional Development: Why is professional development important for team collaboration, and what forms can it take?

Chapter 7: *"Environmental Ergonomics: Optimizing Human Performance"*. These questions are designed to help you reflect on the key concepts of environmental ergonomics and apply them to real-world scenarios.

1. Thermal Comfort: How does thermal sensation affect workplace productivity, and what strategies can be used to achieve thermal equilibrium in an office setting?

2. Heating and Cooling: What are the benefits of efficient heating systems and air conditioning in maintaining thermal comfort, and how do natural ventilation and shading contribute during warmer seasons?

3. Noise Control: How does noise pollution impact concentration and stress levels, and what measures can be taken to mitigate noise in a workspace?

4. Office Acoustics: Discuss the role of sound absorption materials and the use of background noise in creating an acoustically comfortable office environment.

5. Air Quality: Why is indoor air quality important for cognitive performance, and what practices ensure clean air circulation within a workspace?

6. Natural Ventilation: How does natural ventilation affect indoor air quality and the well-being of occupants in a building?

7. Lighting and Circadian Rhythms: Explain the relationship between natural light exposure, artificial lighting, and circadian rhythms in the context of workplace design.

8. Biophilic Design: What is biophilic design, and how does integrating nature into workspaces enhance mood and cognitive performance?

9. Smart Buildings: How do smart buildings utilize sensors to create energy-efficient and user-friendly spaces, and what are the advantages of such systems?

10. Indoor Air Pollutants: Identify common indoor air pollutants and their potential health impacts. How can workplaces minimize exposure to these pollutants?

11. Environmental Ergonomics Applications: Provide examples of how environmental ergonomics principles can be applied to optimize human performance in different types of work environments.

Chapter 8: Self-help questions based on the *chapter "Musculoskeletal Disorders and Injury Prevention"*. Designed to help you reflect on the ergonomic practices discussed in the chapter and apply them to your own work environment to improve safety and well-being.

1. Understanding MSDs: What are musculoskeletal disorders, and which parts of the body do they typically affect?

2. Identifying Risk Factors: Can you list the main risk factors for developing MSDs in the workplace?

3. Ergonomic Workstation Setup: How can you adjust your chair, monitor, keyboard, and mouse to reduce the risk of MSDs?

4. Proper Lifting Techniques: What steps can you take to ensure you are lifting objects safely to prevent injury?

5. Stretching and Exercise: What type of dynamic stretching exercises can you perform to warm up before starting work?

6. Recognizing Symptoms: What are the early signs of MSDs, and why is it important to report them promptly?

7. Seeking Prompt Treatment: Why is early intervention crucial when experiencing symptoms of MSDs?

8. Employee Training: What kind of training should workers receive to minimize the risk of MSDs?

9. Supervisor's Role: As a supervisor, what measures can you take to identify risks and ensure the safety of your team?

10. Common Posture Mistakes: What are some common posture mistakes made while sitting at a desk, and how can they be corrected?

11. Creating a Culture of Safety: How can you contribute to fostering a culture of safety in your workplace?

12. Practical Ergonomic Adjustments: What practical adjustments can you make to your workstation to enhance ergonomics and prevent MSDs?

Chapter 9: *"Ergonomics in Product Design: Creating User-Centric Solutions".* These questions can guide you through the process of creating ergonomic solutions that are both user-centric and innovative. The goal is to enhance the user's experience by making products that are intuitive, comfortable, and safe to use.

1. Understanding Ergonomics

- How can I assess the ergonomic needs of my target user group?

- What methods can I use to ensure my product design accommodates various body sizes and shapes?

2. User-Centered Design

- In what ways can I gather meaningful user feedback during the design process?

- How can anthropometric data influence the design of my product?

3. Key Ergonomic Considerations

- What strategies can I employ to balance aesthetics and functionality in my designs?

- How can I enhance the comfort and safety features of my product?

4. Ergonomic Product Categories

- What are the key ergonomic features to consider when designing furniture for prolonged use?

- How can I ensure handheld devices are comfortable for a wide range of hand sizes?

5. Iterative Design and Prototyping

- How often should I iterate on my designs based on user feedback?

- What are the best practices for prototyping ergonomic products?

6. Future Trends

- How can I incorporate sustainable materials without compromising ergonomic design?

- What are the ergonomic challenges when designing for VR and AR technologies?

7. Poorly Designed Products

- What are common ergonomic design flaws in office chairs, and how can they be addressed?

- How can I redesign a computer keyboard to reduce the risk of RSIs?

Chapter 10: Self-help questions that could be included at the end of the chapter *"Ergonomics in Healthcare and Special Populations: Designing for Comfort, Safety, and Inclusivity"*. These questions are designed to encourage readers to think critically about the application of ergonomics in healthcare and to reflect on how design choices can impact the comfort, safety, and independence of special populations.

Ergonomics in Healthcare Settings:

1. What are the key considerations when designing patient beds to ensure they are adjustable and user-friendly for both patients and caregivers?

2. How can wheelchair accessibility be improved in hospital environments, including corridors, doorways, and restrooms?

3. In what ways can lighting and noise control contribute to patient recovery, and how can they be optimized in hospital settings?

Aging Populations:

4. Discuss the importance of installing grab bars and choosing non-slip flooring in preventing falls among aging populations.

5. What features should be considered when designing assistive devices like canes and walkers to ensure they are lightweight and comfortable?

Individuals with Disabilities:

6. How can workstations be made more accessible for individuals with disabilities, particularly those with limited dexterity?

7. What are the benefits of customizable interfaces, such as adjustable font size and contrast, for users with visual impairments?

Inclusive Design Principles:

8. Explain the concept of universal design and its importance in creating products that are usable by all individuals, regardless of their abilities.

9. How can collaboration with end-users, through co-design and feedback loops, improve the ergonomics of healthcare products and environments?

Future Challenges and Innovations:

10. What role can virtual reality (VR) play in the rehabilitation of patients, and how can these experiences be customized for various conditions?

11. How might smart homes and assistive technologies evolve to better support individuals with mobility challenges and facilitate real-time health monitoring?

Practical Tips and Examples:

12. What are some practical design tips for creating an accessible waiting room in a healthcare facility that ensures comfort, safety, and inclusivity for all visitors?

13. How can clear signage and wayfinding contribute to a more intuitive and user-friendly layout in healthcare settings?

Chapter 11: Self-help questions on *"Enhancing Ergonomics with Exoskeletons."* These questions are designed to prompt critical thinking and deeper understanding of the role of exoskeletons in enhancing ergonomics across various fields.

1. Reducing Physical Strain:

   - How do exoskeletons provide physical support, and in what ways do they reduce the strain on the user's body?

   - Can you identify specific ergonomic problems that exoskeletons can address in reducing workplace injuries?

2. Improving Posture:

   - What features of exoskeletons contribute to maintaining proper posture?

- How does improved posture through the use of exoskeletons reduce the risk of musculoskeletal disorders?

3. Increasing Efficiency:

- In what tasks have exoskeletons been shown to increase efficiency and productivity?

- Discuss the potential impact of exoskeletons on the overall workflow in industrial settings.

4. Applications in Healthcare:

- How are exoskeletons used in rehabilitation to assist patients in regaining mobility?

- What are the benefits and limitations of using exoskeletons in healthcare environments?

5. Military Applications:

- Describe how exoskeletons can enhance the physical capabilities of soldiers.

- What are the considerations for integrating exoskeletons into military operations to reduce fatigue and increase endurance?

6. Challenges in User Comfort:

- What design elements are crucial to ensure that an exoskeleton is comfortable and does not restrict movement?

- How can user feedback be incorporated into the design process to improve comfort?

7. Cost and Accessibility:

- Discuss the barriers to widespread adoption of exoskeleton technology due to cost.

- What strategies could be employed to make exoskeletons more accessible to a broader audience?

8. Regulation and Standards:

- Why are regulatory standards necessary for the safety and effectiveness of exoskeletons?

- What role do you foresee for regulatory bodies in the development and deployment of exoskeleton technologies?

Chapter 12: Self-help questions based on the *chapter "Leveraging AI for Ergonomics"* Designed to help individuals and organizations think critically about the integration of AI in ergonomic practices and to encourage proactive engagement with the technology to enhance workplace safety and comfort.

1. Posture Analysis:

   - How can I use AI to identify and correct poor postures at my workplace?

   - What are the best practices for implementing AI posture analysis systems?

2. Workplace Design:

   - In what ways can AI contribute to more ergonomic workplace designs?

   - How can I ensure that AI recommendations for workplace design are followed?

- Personalized Training:

   - How can AI tailor ergonomic training to my specific needs?

   - What kind of real-time feedback can AI provide to help me improve my ergonomics during training?

3. Product Design:

   - How can predictive modeling with AI assist in creating ergonomic products?

   - What role does AI play in analyzing user testing data for ergonomic design?

4. Challenges and Mitigation:

   - What steps can I take to address data privacy concerns when using AI for ergonomics?

   - How can I recognize and mitigate algorithm bias in AI ergonomic solutions?

Chapter 13: Based on the chapter *"Ergonomics Training,"* here are some self-help questions that could guide individuals and organizations in implementing effective ergonomics training programs: These questions aim to promote a proactive approach to ergonomics training, encouraging continuous learning and adaptation to new technologies and methods for a safer and more efficient workplace.

1. Understanding Ergonomics:

   - What are the fundamental principles of ergonomics that every employee should know?

   - How can I apply the basics of ergonomics to my daily work routine?

2. Identifying Ergonomic Risks:

   - What are common ergonomic risks in my workplace, and how can I spot them?

   - How can I contribute to a safer work environment by identifying and reporting ergonomic risks?

3. Implementing Ergonomic Solutions:

   - What are some practical ergonomic solutions I can implement at my workstation?

   - How can I ensure that the ergonomic solutions are suitable for my specific work tasks?

4. Maintaining Ergonomic Practices:

   - What strategies can I use to maintain ergonomic practices in the long term?

   - How can I encourage my colleagues to continue applying ergonomic principles?

5. Challenges in Ergonomics Training:

   - How can I overcome the lack of awareness about the importance of ergonomics in my organization?

   - What resources can I tap into to provide comprehensive ergonomics training despite budget constraints?

6. Innovative Training Approaches:

   - How can online training platforms be utilized to enhance ergonomics training in my organization?

   - In what ways can virtual and augmented reality contribute to more effective ergonomics training?

7. Technology in Ergonomics Training:

   - How can AI and machine learning be leveraged to personalize ergonomics training?

- What role can exoskeletons and wearable technology play in providing real-time feedback for ergonomic training?

Chapter 14: "Ergonomics and Remote Work." These questions can help guide a discussion or exploration of the importance of ergonomics in remote work settings.

1. What are the key principles of ergonomics that should be considered when setting up a remote workspace?

2. How does proper ergonomic design contribute to productivity and well-being in a remote work environment?

3. What are the recommended heights and distances for computer monitors to reduce eye strain for remote workers?

4. Can you describe the ideal ergonomic posture for sitting at a desk while working remotely?

5. What types of ergonomic equipment are essential for a home office setup?

6. How often should remote workers take breaks, and what type of activities are recommended during these breaks?

7. What role does lighting play in creating an ergonomic remote work environment, and how can it be optimized?

8. How can remote workers ensure they maintain an ergonomic posture throughout the workday?

9. What are some common ergonomic challenges faced by remote workers, and how can they be addressed?

10. In what ways can employers support their employees in achieving an ergonomic setup for remote work?

EPILOGUE

The challenges we've explored in this book—back pain at work, repetitive strain injuries (RSI), and the catastrophic failures of complex systems—are like microcosms within a broader context. These issues are part of a larger domain concerned with the humanization of technology: macro-ergonomics.

Understanding Macro-ergonomics

Macro-ergonomics transcends individual workplaces and tasks. It zooms out to consider entire systems, organizations, and societal impacts. Here are some key points:

1. System-Level Thinking: Macro-ergonomics encourages us to think beyond the immediate work environment. It involves analyzing how various components—people, processes, tools, and technologies—interact on a grand scale.

2. Holistic Solutions: Rather than addressing isolated problems, macro-ergonomics seeks holistic solutions. It acknowledges that changes in one area can ripple through the entire system.

3. Human-Centered Design: At its core, macro-ergonomics aims to create systems that prioritize human well-being. Whether it's designing a factory layout, optimizing transportation networks, or shaping organizational policies, the focus remains on enhancing human performance, safety, and satisfaction.

4. Sociotechnical Systems: Macro-ergonomics recognizes that work systems are not just technical but also social. It considers organizational culture, communication patterns, and the broader societal context.

The Broader Canvas

As we bid farewell to the microcosm of individual ergonomic challenges, let's embrace the canvas of macro-ergonomics. Here, we explore questions like:

- How can we create workplaces that foster collaboration and innovation?

- What policies promote work-life balance and employee well-being?

- How do we design transportation systems that minimize stress and maximize efficiency?

- Can technology enhance—not hinder—our quality of life?

Remember, the ergonomic journey doesn't end here. It extends beyond cubicles and assembly lines. It touches our homes, cities, and global networks. So, as practitioners, researchers, and advocates, let's continue shaping a world where technology serves humanity, not the other way around. Thank you for joining me on this enlightening voyage. May your ergonomic endeavors be both impactful and compassionate.

—Paul Jos

BIBLIOGRAPHY

1. Stack Theresa, Lee T. Ostrom. Occupational *Ergonomics: A Practical Approach*. 2nd Edition [Wiley]. Publication Date: 08/11/2023

2. Palmer, Pepper. *Ergonomics: Improving System Performance and Human Well-Being.* [Willford Press]. Publication Date: September 26, 2023 (source; amazon)

3. Guastello, Stephen J. *Human Factors Engineering and Ergonomics: A Systems Approach.* 3rd Edition. [CRC Press]. Publication Date: April 14, 2023 (source; amazon)

4. Ortiz Hugues, Juan Carlos. *Ergonomics Applied to Dental Practice.* [Quintessence Pub Co]. Publication Date: January 1, 2023 (source, Amazon)

5. Openshaw, Scott, Taylor, Erin. *Ergonomics and Design: A Reference Guide.* [Allsteel Inc]. Publication date: 2006

6. Alvin R. Tilley, Henry Dreyfuss Associates. *The Measure of Man and Woman: Human Factors in Design.* Revised Edition [Wiley]. Publication Date: December 31, 2001 (source, Amazon)

7. Bridger, Robert. *Introduction to Ergonomics, 3rd Edition.* [CRC Press] Aug 14, 2008. (sources google books)

8. Salvendy, G. *Handbook of Human Factors and Ergonomics Methods.* 4th Edition [Wiley] Publication Date: 2012.

Nb: Beyond the textbooks and reference books cited in the bibliography, this work has been enriched by a broad spectrum of resources. These resources, which have contributed to the depth and breadth of the content, extend beyond the conventional written form, encompassing both style and substance. They include academic journal articles, conference proceedings, theses, and dissertations, as well as government documents. Personal interviews and articles from newspapers and magazines, both print and online, were also consulted. Additionally, websites, online resources, and multimedia sources such as podcasts, videos, and broadcasts were utilized. It is of paramount importance to acknowledge all these sources, as each has played a significant role in shaping the research and influencing the ideas presented in this work.

Paul Jos - I also blogging at: https://beyourownsafetychampion.blogspot.com/

www.ingramcontent.com/pod-product-compliance
Lightning Source LLC
Chambersburg PA
CBHW062117220526
45471CB00010B/3775